alue Analysis: Way to Better oducts and Profits

David Fram

AMA management briefing

Value Analysis: A Way to Better Products and Profits

Value Analysis: A Way to Better Products and Profits

an *AMA* management briefing

David Fram

amacom
A DIVISION OF AMERICAN MANAGEMENT ASSOCIATIONS

 Printed in the United States of America.

This Management Briefing has been distributed to all members enrolled in the Manufacturing Division of the American Management Associations. A limited supply of extra copies is available at $5.00 a copy for AMA members, $7.50 for nonmembers.

International standard book number: 0-8144-2172-5

First printing

Contents

David Fram joined Sperry Rand Corporation in 1937 as a tester of aircraft flight instruments. He remained with the corporation until his retirement in 1974. During that time he held various positions in quality control, engineering, manufacturing, and works management. In recent years he served as corporate training coordinator with the responsibility for introducing value analysis techniques and programs in the United States and Europe. During 1973 he served on the Economic Development Council of New York City in a task force effort to reduce the administrative burdens of high school principals. He is a frequent lecturer for the American Management Associations and a value analysis consultant, located in Great Neck, New York. Mr. Fram holds B.S. and M.S. degrees in physics from Worcester Polytechnic Institute.

I

What Is Value Analysis All About?

The emphasis today is on improvements that can make for better products and services, and for improved returns on investment. Value analysis is one management tool that can be applied for this purpose. Basically it is an analytical approach for examining the relationships between the performance, or satisfaction, provided by a product or service, and the cost of making the *essential* performance or satisfaction available. Such an examination goes well beyond the usual cost reduction methods, which focus on intermediate economies. Value analysis, for example, might suggest that the manufacture of a particular product be discontinued in favor of another that could more economically provide the *essential* functions of the first one.

There are many avenues open to a management that is seriously concerned about improving profits and return on investment. One of these is an industrial technique developed by Lawrence D. Miles, formerly of the General Electric Company, back in the 1940s. It deals with the concept of value—giving customers better values and suppliers satisfactory profits. During the past 20 years, this technique has become known by many names, including value analysis, value engineering, value control, value assurance, and value improvement. For our purposes we shall consider all these titles as referring to the same technique. The important word here is "value."

What is value? Unfortunately, value is an abstract concept with a variety of meanings. If we turn to the dictionary, we find that "value" is defined as the "worth of a thing, utility, the ratio of utility to price, a bargain," etc., etc. In some industrial situations "value" means long life, high precision, good performance, or, ultimately, low cost. None of these definitions helps us very much to improve value in a systematic way. As a result it is necessary to develop a definition of value that can guide an organized effort toward significant improvement: "Value is the relative cost of providing a necessary function or service at the desired time and place with the essential quality."

It is important to note that in this definition value is a relative matter—the result of comparing one thing with another. Next, note that value deals with *necessary* functions or services, which in the final analysis are what the customer is really buying. Value is also directly related to schedule. If we can't procure what we need when we need it, value sometimes decreases significantly. Note, too, that quality is a variable factor, depending on customer needs and expectations, and that this provides a base line to which other values can be related. In this view, value is not an inherent characteristic of an item; value depends on specific, identified, and defined factors and features.

Value analysis employs a new way of thinking about value. We start with the premise that value is relative—the result of comparing the cost of providing a necessary function by one means, with the cost of providing the same function in a different manner. Value analysis is concerned with identifying the unnecessary costs that do not add essential reliability or quality to the product.

Quality, for our purpose, depends on the capability of a product to perform its required functions. Value analysis examines the value of a product in terms of its performance of designed functions at lower cost. Quality control and value analysis have similar, though not identical, objectives. Value analysis attempts to reduce costs while maintaining quality, functional performance, and customer goodwill at acceptable levels.

There are several types of values. *Exchange value* results from the characteristics of an item that make it possible to exchange it for something else; example, paper money. *Esteem value* refers to the features of an item that motivate us to want it; example, the vinyl top on a sedan automobile. *Use value* refers to those qualities of an item that accomplish a desired function; example, the warmth provided by an electric heater. Other types of value might be classified as sentimental, religious, and historical.

Value analysis, as discussed here, will for the most part deal with use value. We are concerned primarily with how to perform a necessary function at the lowest cost consistent with the quality level desired.

2

Value Analysis and Cost Reduction

At first glance value analysis might seem like nothing more than good old-fashioned cost reduction. But this is not quite true. Cost reduction should be a part-time activity for everyone in an organization, and it should be continued. Value analysis is not designed or intended to replace it. Consider this example: One of the components in an electronic music organ is composed of 163 parts. Its function is to produce a vibration in the tone similar to what the violinist does when he rocks his finger on the violin string. The unit performed its assigned function very well. Each of the 163 parts was manufactured carefully, accurately, and at the lowest cost, in accordance with good cost reduction techniques. In addition, those items that were procured from suppliers were purchased at the lowest competitive prices. However, when this design was analyzed for value, the same function was achieved with a unit that had only 51 parts. The improvement required a new way of thinking about the product and the characteristics that gave it value.

Decision makers typically operate under intense time pressures. As a result, subsequent decisions usually include improvable elements, which are discovered when the elements are examined at greater length. The original design made it possible to produce the product within a specified time frame; that is, according to schedule. No criticism of the original designer should be implied from a second, more leisurely look resulting in better value. The second look probably would not have been taken in the absence of the first one. In many instances, third and fourth looks—or more—will continue to produce additional value. The improvement of value is a never ending process, and the contributions of all participants should be appropriately recognized.

Why is value analysis important? There are two reasons. First, if company A has a well-organized value analysis program and its competitor, B, does not, it is just a matter of time before company A outperforms company B. In time, company B may have to close its doors permanently. Second, many studies have shown that every industrial organization operates with unnecessary costs, and these costs reduce profits.

Beyond benefits to the individual company, value analysis ultimately benefits customers by providing them with lower-cost methods of satisfying their functional needs. In the long run this makes a significant contribution to a rising standard of living and diminishing demand on critical resources.

3

How Extra Costs Creep In

There are many reasons why unnecessary costs are a part of our normal industrial operations. Here are a few:

Flood of information. Although the amount of information available today in every industry is voluminous, it is extremely difficult for any individual or group of individuals to keep up with it. However, unless all the information *relevant* to a value decision is collected and considered, unnecessary costs may result from judgments based on incomplete data.

Habitual thinking. Most of us, particularly when operating under time pressures, tend to rely on habit in making simple decisions. In our private lives we fall back on habit in much of what we do. That's OK, but if we apply habitual thinking to our business operations we are inviting extra costs.

Negative attitudes. The fear of failure causes many people to respond negatively to proposed changes. Often the more authority that is associated with the position, the greater is the tendency to reject new ideas. Yet the acceptance of new ideas and the implementation of changes are probably the most productive approaches to eliminating unnecessary costs.

Reluctance to seek advice. As people acquire more responsibility and authority they often develop the feeling that their success is due entirely to their own knowledge and skill. To ask for help in a problem situation is then considered a sign of weakness, to be avoided at all costs. One result is certain—unnecessary costs.

"Comfortable" high-cost activities. In most established organizations there are activities that function very well but cost too much. Because these things have existed for a long time, and the amount of money involved is usually small, nobody wants to, or has the time to, bother with them. As a result these unnecessary costs continue year after year. In one case, for example, a small electric motor was used to drive a variable condenser through a specially designed coupling made of a stainless steel bellows. After value analysis, it was replaced by a piece of plastic tubing, with a 95 percent reduction in cost.

4

The Nature of Value Analysis

Value analysis must be understood, accepted, and supported by top management. Value analysis proposals require that something be changed in order to produce improvement, but experience shows that all levels of management tend to resist change simply because it disturbs the known and familiar. Human nature, like everything else in the universe, if at rest tends to remain at rest. Value analysis is usually administered as a staff function. To achieve any significant improvement it must be accepted and implemented by line personnel, and this will be done only if the activity is strongly supported by top management.

Value analysis is a group activity. It is a proven fact that a group can create more good ideas than individuals working alone. Many organizations use value teams. A value team is a small, carefully selected group of individuals organized to analyze a specific item in accordance with a planned schedule. Normally, the value team should contain individuals from engineering, manufacturing, quality control, and purchasing. In some cases it should also include individuals from marketing, field service, and accounting. A department usually has a sense of identity and pride in its particular activity. It also tends to resent another department's telling its people what to do or how to do it. Thus the value team should include representatives from all departments that will participate in implementing the changes.

The value analysis concept also has marketing implications. The desirability of bringing vendors into planning and value analysis activities is a useful initiative for the organization engaged in value analysis. The concept can also be applied by the vendor organization, on its initiative, in exploring either alone or in conjunction with a customer organization the various ways in which the customer's functional needs might be met. Thus the value of the vendor's product or service will be enhanced, which in turn provides a better selling opportunity when the customer recognizes that a proposed innovation represents a better value. The result is that the profitability of both organizations, and the value provided the ultimate consumer, is improved.

In this connection it is quite possible to operate on several value levels, each one matched to the functional needs and value judgments of a particular market segment. This is done, for example, by automobile manufacturers in supplying their customers with functional transportation in varying configurations of cost, luxury, and performance.

Value analysis can properly emphasize vendor assistance. Every company is consistently faced with make-or-buy questions. Twenty-five years ago the generally accepted policy seemed to be to "make it in house if at all possible." Today, although there is still a strong trend toward in-house operations, many managements recognize that they can't do it all at the lowest cost. Suppliers who can deliver quality on schedule are being called in, early in the design stage, to assist in arriving at better values. Suppliers have a great deal of knowledge about some products. And suppliers have the advantage of looking at a situation from the outside—the trees don't prevent them from seeing the forest. For best results, of course, the supplier will need to have adequate information concerning the end product and how it will be used.

5

Operating the Value Analysis Plan

Value analysis is best practiced in accordance with a specific plan, which consists of five steps:

1. Information phase
2. Speculation phase
3. Analysis phase
4. Execution phase
5. Reporting phase

Let's take a close look at each phase.

Information Phase

In the first phase of the plan, we ask three simple questions:

1. *What is it?* This question calls for a complete definition of the item being subjected to a value study. Such information as the name of the item, any drawing numbers involved, where the item is manufactured, what materials are used, how it is made, what the associated parts are, what quantity is needed, and any special requirements should be obtained.

2. *What does it cost?* It is very important that this answer be as accurate as possible, but if actual figures aren't available, a good estimate will suffice.

3. *What does it do?* Here we are dealing with the most basic element of the entire value analysis technique. The ultimate value of an item depends on how well it performs the function for which it was designed. It is not unusual for an item to perform several functions. Each function must be carefully and completely defined.

But what do we mean by "function"? One dictionary states that function is: "the specific, natural, or proper action that belongs to an agent; the normal action of any organ or set of organs." These words don't help the value team very much in the development of a working definition of an item that can be used to improve the value of that item. For value study purposes a much better definition of "function" is: "Function is what makes the product, process, or procedure work or sell."

We may consider that there are two types of functions: basic, or primary, and secondary. A basic function is the indispensable, or essential, one for which a device was designed and manufactured. Secondary functions are those that are subordinate or supplemental to the basic function. For example, if we ask what the function of a steel drill is, one answer would be that it drills holes. This is not the best definition because it restricts us to drilling as a way of making holes. A better definition might be "removes material" because it is essential that we consider other methods of making holes besides using a drill. Similarly, if we ask what the function of a power cable is, a quick answer might be that it conducts electricity. Here again, a better definition of function is "transmits energy" since we should consider other methods of achieving the desired function without depending on electricity. Let us now apply these ideas concerning function to an ordinary flashlight and its component parts (Table 1).

From the table it is quite obvious that only five components of the flashlight are involved in the basic function; the other parts perform only secondary functions. One manufacturer of flashlights, aware of this situation, produced a flashlight in which the rear cap and threads on the rear cap and body were incorporated in a front-loading plastic body. The result was a flashlight that provided the basic function at a lower cost.

The principal advantage of this analysis is that it distinguishes the parts that are involved in basic functions from the parts that perform only secondary functions. Careful analysis of the parts that perform only secondary functions often leads to combinations and eliminations without downgrading the overall quality of the equipment. Gathering all the information for a value study and defining all functions is not simple. Too often it is treated much too lightly, and the results are far less than they should be. Information must be obtained from the best sources, both within the organization and from outside sources. A value team's success largely depends on how thoroughly it has built up an information base from which to move into the speculation phase.

Speculation Phase

Here we ask, "What else will do the job?" This question calls for creative thinking, a process that is not quite as mysterious as we sometimes suppose. Consider the kaleidoscope an analogy to creative thinking. The more pieces of colored glass there are in the drum, the more patterns we can produce. In a similar manner the more bits and pieces of knowledge we have in our minds, the more patterns, combinations, or ideas we can produce. However, the presence of the bits and pieces of knowledge in our minds doesn't assure us that different ideas will result. In the case of the kaleidoscope, it is necessary to rotate the drum; in our minds we must arrange, and rearrange, our knowledge in order to create new ideas. We can state this concept in the form of an equation:

$$\text{Creativity} = \text{knowledge} \times \text{imagination} \times \text{evaluation}$$

If knowledge is zero, any amount of imagination and ability to evaluate will result in zero creativity. Similarly, all the knowledge in the world and zero imagination or zero evaluation produce zero creativity. The important conclusion we can draw from this analysis is that is behooves each individual to maximize

TABLE 1

Item	*Function*	*Basic*
Flashlight	Provides light	X
Components		
Lens	Protects bulb	X
Reflector	Focuses light	
Rear cap	Retains spring	
Threads on rear cap and body	Permits access	
Bulb	Provides light	X
Front cap	Holds lens	
Threads on front cap and body	Permits access	
Battery	Provides energy	X
Spring in rear cap	Provides tension	
Brass strip in body	Transmits energy	X
Body	Holds parts	
Switch	Connects/disconnects energy	X

his knowledge, imagination, and creative ability in order to achieve maximum creativity.

We all know that some people are more creative than others. It is also true, however, that a lot of the difference is the result of varying degrees of energy and motivation rather than inborn talent. And, of course, creativity is not encouraged when the boss starts a meeting by saying, in effect, "All those in favor say, 'Aye'—those opposed say, 'I quit.'" Other deterrents to creativity are conformity, fear of failure, superperfectionism, and just plain laziness. When we fall back on habit, simply because we are "too busy" to think, product value is adversely affected. A good piece of advice in this connection is: After doing something the same way for two years, look it over carefully; after five years, become suspicious; after ten years, throw it out and start all over again.

In the exotic space age we live in, the mention of problem solving quickly brings to mind the computer, a machine that can add, subtract, multiply, divide, compare, and select in millionths of a second. But so far, no computer produced by humans is able to create. The only source of creative thinking available to mankind is the human brain. Recent studies show that it contains about 13 billion nerve cells. About one-third are used during a normal lifetime, only 5 percent are in constant use. If we could find some way to increase that one-third to one-half, and bring the 5 percent up to 10 percent, we would find it much easier to solve problems and make decisions. That increased utilization is what creativity is concerned with. It is not something possessed by a few special people. Psychologists working in the field have concluded that everybody has some creative ability, and the records show that many ordinary people have at times demonstrated unusual creative talent. All the studies suggest that almost everyone can become more creative. The individuals who work at it can become more creative. There is no guarantee that they will become Einsteins or Edisons, but there is every reason to believe that they can become much more creative than they now are.

Problem-solving Sessions

The basic rules for conducting a problem-solving session were developed by Alex F. Osborn of the Creative Education Foundation in Buffalo, N.Y. Paraphrased, they are:

> When people are thinking up new ideas, criticism should be ruled out. Don't evaluate ideas while they are being created; let the ideas flow until every alternate is available for consideration. The process of holding back criticism can be compared with driving down an open thruway. The driver can travel at high speed and cover a great distance in a short time because there are no intersections or opposing traffic. The driver can reach his planned destination quickly and easily. What happens when a stop sign suddenly appears? The normal instinct of a good driver is to bring the vehicle to a dead stop. For the moment the driver is going nowhere. Criticism is similar to a stop sign—it brings creativity to a sudden stop. Everyone immediately gets involved in criticizing the new idea and in defending his own. When you apply the brake, the vehicle stops; when you criticize, creativity stops.

When we read, observe, or listen we increase the pieces in the kaleidoscope; that is, we increase our ability to form new patterns or, in value analysis terms, new ways to perform necessary functions. In a group creativity session individuals can exchange ideas, with the result that everyone has an opportunity to make a contribution toward solving the problem.

Separating the creation of ideas from their evaluation is called deferred judgment. The principle of deferred judgment has been tested many times. In most cases the teams that used the principle produced a lot more good ideas. The basic concept is to be a thinker when trying to create new ideas, and to be a judge later, when ideas are evaluated. The secret of success is to keep the two processes separate.

Freewheeling should be welcomed. This

means the wilder the idea, the better. Ideas that may seem ridiculous at first after a little thought often lead to sound, practical alternates. To really make this concept work, one must take a "shoot for the moon" approach. Forget logic, experience—all the limitations that surround us. Think about the unusual, the unfamiliar, or what is different but in some ways related to what is being studied. Give your imagination free rein—let it go wherever it will. In a creativity session the sensible, practical ideas come to the surface easily. After these stop flowing, it is time to go after the really far-flung, imaginative, and creative ones.

In this phase of the plan we recommend the technique "Blast, create, and refine." To use this technique successfully, it is necessary to blast through regimented ways of thinking, find new ideas, and refine the raw ideas by using sound judgment to obtain the desired functions and essential reliability.

This way of thinking is designed to take us beyond the conventional or 10 percent cost reduction that is normal in most industries. If we are motivated to move into new areas—areas that have never been considered before—it is sometimes possible to reduce costs by as much as 75 to 90 percent. Thomas Edison tried more than a thousand kinds of fiber before finding a filament for his lamp. It has been said that man's mind is like a parachute: to work, it must first be open.

Recently, a value team was looking at a sketch of an airplane radar set cover. It was 24″ in diameter, 30″ deep, and made of aluminum; it would cost $90. Its function was to protect equipment. During the brainstorming session, various other ways of performing the function were considered. One member of the team mentioned that a wastebasket had the general shape of the radar set cover and cost only about $1.50. Another member pointed out that a garbage can cost only about $3. A restaurant-size cooking pot was discussed, and finally a fiberglass planter.

It didn't take very long for the team to select a fiberglass part that would cost only $5. The freewheeling brainstorming effort resulted in a part having a much higher value than the original aluminum spinning. But something else important happened here. When was the last time you were in a problem-solving session and someone had the courage to say, "Let's consider a garbage can as an alternate?" That individual most likely would have been laughed out of the room. Unless the working atmosphere in an organization is such that every individual feels free to suggest that an item under consideration is similar to a garbage can, that organization is not operating at its highest creative potential, and its products probably do not represent the best value for its customers.

There are no bad ideas, only good ones and better ones. Most brainstorming sessions get bogged down in yes-and-no arguments over suggestions to change. Consequently many of the best ideas don't get sufficient consideration. There is a fundamental question that each individual involved in a brainstorming session should ask himself before participating. "Am I the kind of person who responds first to what is good in an idea, or do I respond first to what is bad?" Obviously a value team composed of "yes" members can be much more creative than a team consisting of "no" members, and their efforts will consistently produce solutions of higher value. If all participants will practice deferred judgment, it is possible to avoid losing good ideas in a flood of criticism.

Quantity is needed. The project team must drive itself to create as many ideas as possible. Everything that comes to mind should be written down. The more ideas available to the team, the greater the probability that the team will have more good ideas for final consideration. Experience shows that the best ideas do not appear at the beginning of a brainstorming session. An in-depth study of this activity showed that the second half of the session produced about 80 percent more good ideas than did the first half.

Combinations and improvements are sought. Try combining two or more ideas into a new idea. Improve old ideas and methods. Consider both major and minor variations of what is now being done. *Substitute* other parts, other materials, other processes or procedures. Example: the channel material in a computer tray was replaced by formed wire at 40 percent less cost. *Reverse* things or procedures to achieve better value. Example: a cooling circuit was required in a beryllium copper mold for a plastic process. The normal approach would be to drill holes in the walls of the mold. Better value was achieved by starting with a ¼″ copper tube placed in a Shaw process mold and pouring the beryllium copper around it. *Combine* old ideas into a new idea—use blends, alloys, or mixtures. Example: a relatively new product on the market is a self-tapping screw, which is simply a combination of a tap and a screw that performs both functions in the same item. *Develop new uses.* Example: a heavy-duty helicopter can be used to deliver a completed structure to a building site. *Make it smaller.* Example: A stapling mechanism no bigger than a fountain pen. *Make it bigger.* Example: A dump truck with a capacity of 200 tons, 40 mph speed, 1600 horsepower, and a weight of 650,000 pounds.

We all come into the world as infants with imaginations that are active throughout childhood. But during the teenage period it begins to decrease. Imagination continues to decrease in early adulthood because we start to develop judgment. As the judgment factor increases, the imagination factor decreases and sometime during middle age, they cross. When we reach old age, too many of us have developed judgments that overpower whatever little imagination has survived the years. To combat this trend, every individual should make every effort to keep both imagination and judgment active.

Analysis Phase

Here we ask: Will it work? What will it cost? This phase should be completely separated from the speculation phase. In some situations it is best done by a separate group to preserve objectivity. The first step is to eliminate ideas that obviously don't apply to the situation at hand. Succeeding steps involve evaluation of all ideas that represent possible solutions. It is often helpful to discuss some matters with experts within or outside the company. As the number of alternatives is reduced it becomes very important that reliable cost estimates be prepared. It is also important to keep in mind the distinctive performance characteristics the company emphasizes, whether it be long life, precision, high performance, low cost, or a combination of these factors.

This is the time for rigorous thinking, critical evaluation, and the use of all the resources of the organization. However, it is often possible, when all functional points of view are represented on the team, to reach a conclusion by carefully comparing the ideas that survive the mutual screening. In any event, the important thing is to prove that the proposed product or service will provide better value than what it would replace. This may require experimenting or building a model.

Experience has shown conclusively that team members assigned to value analysis should not be those who would have the authority and responsibility for implementing the proposal. Their presence has almost always acted as a deterrent to creative approaches. Consequently it is now necessary to locate the individual or individuals who have the authority to pass final judgment on the recommendation and get it implemented. Here top management support is again essential. All the effort expended before this phase may go down the drain unless the individual with the authority to accept or reject the value proposal can find the time to do the job. Nothing can kill a value program faster than to have good proposals end up in someone's bottom drawer.

Execution Phase

It is a fact that most people resist change. History is replete with such incidents. A committee to consider the telephone as an investment is said to have made made the following report:

1. The telephone is so named by its inventor Mr. A. G. Bell, who sees for it a vast future

as a means of personal communication by voice. He believes that one day a telephone will be installed in every residence and place of business.

2. "We note that Mr. Bell's profession is that of a voice teacher and, particularly, a teacher of the deaf. He appears to have no direct experience with the telephone or any other form of communication, electrical or otherwise. Yet he claims to have discovered an instrument of great practical value in communication, which has been overlooked by the thousands of workers who have spent years in this field.
3. "Mr. Bell's proposal to place his instrument in almost every home and business house is fantastic in view of the capital costs of installing the endless numbers or wires and cables that would be demanded. The central exchanges alone would represent a huge outlay in real estate and buildings, to say nothing of the electrical equipment.
4. "Mr. Bell expects that the public will use his instruments without the aid of trained operators. Any telegraph engineer will at once see the fallacy in this plan. The public simply cannot be trusted to handle technical communications equipment. In any home where there are children, to mention only one point, there would inevitably be a high rate of breakage and frivolous use of the instruments.
5. "While every telegram constitutes in itself a written record of what has been communicated, Mr. Bell's instrument uses nothing but the voice, which cannot be captured in any concrete form; therefore there would be no record of what was said or agreed upon. We leave it to you to judge whether any sensible man of business would transact his affairs by such a means of communication.
6. "Mr. Bell expects that the subscribers to this service will pay to have the instruments installed in their premises and will thereafter pay for each call made. We feel it very unlikely that any substantial number of people will agree to such an arrangement, in view of the telegraph offices which are now giving efficient round-the-clock service in every neighborhood and in the smallest towns, which charge only for actual messages sent according to length.
7. "In conclusion, this committee feels it must advise against any investment whatever in Mr. Bell's scheme. We do not doubt it will find a few uses in special circumstances, such as between the bridge of a ship and the engine rooms, but any development of the kind and scale which Mr. Bell so fondly imagines is utterly out of the question."

Everyone associated with value analysis should realize that resistance to change is a basic human characteristic. If value analysis is to be successful, some way must be found to overcome this resistance, since value analysis is always directly involved with change. Note that much of this resistance may be due to the way changes are made. Recent experience with value analysis shows that when people participate in creating change, a very large part of the rejection problem disappears. This is one reason why value analysis should be practiced as a group effort.

Here are some practical suggestions:

1. Concentrate on specifics.
2. Spend company money as prudently as you would your own.
3. Consider manufacturing ease and the possibility that the product may be over-engineered; for example, with respect to materials, processes, or tolerance.
4. Seek constructive expert advice.
5. Avoid mental roadblocks, limiting habitual thinking and negative attitudes.
6. Realize that there is no payoff unless the good ideas are implemented.

Reporting Phase

This phase involves all the communications problems inherent in an organization, and deserves special attention. A value proposal requires careful planning to insure acceptance and implementation. The basic elements are—

1. A complete explanation of existing methods, including detailed cost figures.

2. A concise definition of basic functions being performed.
3. A detailed description of an alternate method for performing the functions and providing better value.
4. An analysis of the features of the alternate method that make for better value.
5. An estimate of the dollar costs of the alternate method and of the savings to be expected.

Here is where top management plays an important role. It is the responsibility of top management, having decided it wants a successful value program, to support it on a long-term basis. It must communicate its conviction that a value proposal is not a criticism of anyone associated with the original development. If this policy isn't followed in all circumstances, at all times, value analysis will be ineffective and the company would be better off without it.

Avoiding Pitfalls

To increase the probability of the success of a newly launched value analysis program, some commonly occurring pitfalls should be recognized and avoided.

1. Don't rush in to reduce the cost of large, complex assemblies. Like other human skills, functional thinking requires training and practice—we should learn to walk before we run.
2. Don't assign individuals who are not fully convinced that value analysis can actually improve results. People who are unable to accept the concepts of value analysis can be an insurmountable roadblock to progress.
3. Proposals for improvements are sometimes interpreted as personal criticism by those associated with existing designs and procedures. This attitude can be a serious deterrent to achieving significant improvement through value analysis, and it is aggravated when management pointedly asks, "Why didn't we do it this way in the first place?"
4. Don't propose changes that have been insufficiently developed and analyzed. In the early stages of a value analysis program, there is a tendency to latch onto the first idea that looks good and present it for implementation without adequate consideration of alternates, and without an in-depth analysis. Developing alternative ideas is not easy for many individuals, and in the beginning more time will be required for this purpose.
5. Don't believe that value analysis can be applied to an item only once. Experience shows that many items can be successfully re-analyzed. The constant introduction of new materials and new processes provides continuous opportunities for further improvement.
6. If the words "value" and "value analysis" seem to cause a disproportionate amount of resistance in your organization, try using other terminology. There are many well-organized, successful programs that use all the concepts and techniques of value analysis without using the word "value."

6

The Potential of Value Analysis

Where can value analysis be applied? Almost anything that costs money in a company can be the subject of a value study—specifically, items purchased in large quantities, items where product design is pushing the state of the art, highly complex items, high-cost items, materials, parts and components, manufacturing processes, tools and equipment, supplies and services, systems and procedures, and overhead items. There should be no untouchable areas for value analysis.

Value analysis may be organized in many different ways, depending on the size and type of company and the kind of product or service provided its customers. The introduction of a value analysis program in a company should start with an orientation seminar for management personnel. The seminar may last a full day or as long as a week. The objective is to expose managers to value analysis concepts and allow them to apply these concepts to simple items that are a normal part of the company's activities. The workshop part of the seminar should be designed to convince the staff that the concepts are real and when properly applied can produce better values.

After top management and the operating staff are convinced that a formal analysis program will benefit the company, a director should be selected. The qualifications for filling this assignment are rather special. The individual selected must thoroughly understand the concepts and techniques of value analysis and must be convinced that they can be applied to the company's products to achieve better value. He must be mature, self-confident, and enthusiastic. He should have an open mind and respond positively to a new idea. His judgment should be balanced by his imagination.

Value analysis operates best as a staff function particularly because it must work across departmental lines. The man in charge of value analysis should feel comfortable in a staff position and recognize its limitations. He must be a capable leader and be able to get people with different backgrounds and interests to work together effectively.

Where should value analysis report in the organization? This is a very difficult question because organizations vary a great deal in structure. In one company it may be marketing, in another it may be engineering or manufacturing. As noted earlier, the project team assigned to a value improvement study should contain members from each of the major departments that make value decisions, including engineering, manufacturing, quality control, and purchasing. When the situation warrants, members from marketing and finance should also participate. Theoretically, the function should report to top management, in part because that reporting line offers the best opportunity for top management involvement and most effectively demonstrates top management support to the rest of the organization. In practice it usually is attached to the department with the greatest clout in the organization, but if this restricts freedom of movement the potential for maximum benefit will be reduced accordingly.

Management should assume responsibility for innovation. While in the past innovation was a once-in-a-while, catch-as-can operation, proper use of value analysis makes it possible to innovate in an organized, systematic manner with greater probability of success. Progress results from useful innovation, and top management has a responsibility for creating and maintaining an atmosphere that supports innovative, imaginative approaches to improved values.

Value analysis provides management with a proven and organized technique for stimulating and implementing innovative changes. Where it is used as an element of a company's operations and receives the full support and cooperation of all concerned personnel, the potential for improvement is almost unlimited.

The initial result of a successful value analysis program is likely to be a more profitable product or service. The ultimate result will probably be a better value for the customer. If a portion of the improved value is passed along to the customer, the market for the product may be enlarged, and this in turn may further improve the profitability of the originating company. A better-value product can mean better field performance, easier manufacturing, less capital tied up in equipment or inventory, less direct or indirect labor, and so forth. Other benefits can result from the additional consideration given make-or-buy decisions and the incorporation of vendor know-how into product design.

For the company that initiates and implements a successful value analysis program the result usually is increased profitability and growth. For society at large the long-term effect is that better values are made available with reduced demands on energy and materials. In view of current and prospective shortages, these reductions in resource consumption can be expected to become increasingly more important, and even critical, factors in corporate success.

7

How to Get Started

Value analysis involves a new way of thinking about value, cost, function, and worth. To obtain the benefits of this way of thinking, management and decision-making line personnel will need to understand the concepts and be able to apply them to day-to-day operations.

Reasonable competence in value analysis may be acquired in many ways. In addition to seminars offered by management associations, there is a growing supply of formal literature that can be used to acquire an understanding of concepts and techniques, and there are consulting organizations that specialize in value analysis training. Following management's initial decision to use value analysis, a program should be undertaken to give key personnel a working knowledge of value analysis and its application. A program might, for example, include the following steps:

1. Top management makes a clear and visible decision that a value analysis program will be installed.
2. Middle management is brought into the picture and given an explanation of the program and its objectives.
3. At this point it may be helpful for management to bring in a consultant to conduct a short explanatory seminar for middle management personnel. This seminar should establish a knowledge base for people who will be called on to participate in the program. In addition, it should encourage the discussion of all phases of an operating value analysis program and answer questions that might otherwise block an effective effort.
4. An individual from within the organization, or one employed especially for the purpose, should be given the task of becoming thoroughly familiar with value analysis so that he can train others who will be called on to participate in value studies. An alternate approach is to employ a consultant organization to conduct training seminars, which typically will take a full five-day period.
5. The managerial responsibility for the value analysis effort must be assigned by top management.
6. The assigned individual now has the task of getting the program out of the talking stage and into the operating stage. A recommended sequence is:
   (a) High cost areas that lend themselves readily to a value study should be identified. The first items for study should be simple rather than complex.
   (b) A value team of about four members should be organized from those who have attended a training seminar and

are available to spend at least a half day a week on a value study assignment.

(c) The team should make a detailed functional analysis of identified items, using the concepts and techniques of value analysis to develop a lower cost method of performing the necessary functions.

(d) The proposals should be carefully and fully documented and presented to the individual having managerial responsibility for the program.

(e) If the proposal is accepted, top management should take advantage of this opportunity to recognize the contribution of the team members, and to publicize the resulting benefits to the company. This will encourage others to participate in future studies.

(f) A complete accounting should be kept of all costs incurred in carrying out the program and of all savings realized from implemented changes. The ultimate effectiveness of a value analysis program is measured by the net savings—the total savings, less costs incurred.

Good results from a value analysis program relate to both the technique of value analysis itself and the motivation of the individuals associated with it. The effectiveness of the program will depend a great deal on the way it is introduced and administered. A strong showing of support from top management, broad participation in planning, continual managerial reassurance that original working methods and materials are not under criticism, and well-publicized recognition of the benefits that have been realized through the program—all will help make the program successful and profitable.

Suggested Readings

Considerable intellectual stimulation relevant to the tasks and opportunities of value analysis will be found in the magazines listed below.

The Journal of Creative Behavior. Published quarterly by the Creative Education Foundation, Inc. Address inquiries to the managing editor at State University College at Buffalo, 1300 Elmwood Avenue, Buffalo, N.Y. 14222.

Performance. The official magazine, published bimonthly, for the American Society for Performance Improvement, Society of American Value Engineers, National Association of Suggestion Systems, and National Property Management Association. Address inquiries to Charger Productions Incorporated, 34249 Camino Capistrano, Capistrano Beach, Calif. 92624.